YOUR KNOWLEDGE HAS VALUE

- We will publish your bachelor's and master's thesis, essays and papers

- Your own eBook and book - sold worldwide in all relevant shops

- Earn money with each sale

Upload your text at www.GRIN.com and publish for free

Bibliographic information published by the German National Library:

The German National Library lists this publication in the National Bibliography; detailed bibliographic data are available on the Internet at http://dnb.dnb.de .

Imprint:

Copyright © 2019 GRIN Verlag
Print and binding: Books on Demand GmbH, Norderstedt Germany
ISBN: 9783668939516

This book at GRIN:

https://www.grin.com/document/470964

Daniel Stark

A New Concept for Mass

GRIN Verlag

A New Concept for Mass

Daniel Lee Stark

March 9, 2019

Abstract

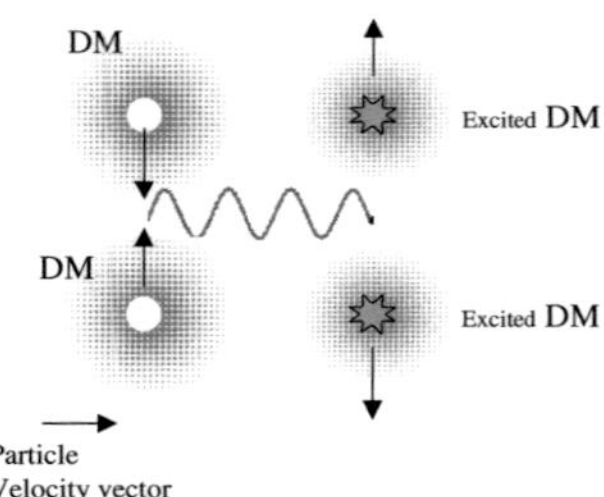

This paper proposes a model for mass that combines two theories.

Gunnar Nordström's theorized mass is dependent on the scalar gravitational field[1] and Stark theorized that our elementary particles are composed of fundamental excited Dark Matter (DM) particles.[2]

Dm excitation theory is that our elementary particles are DM particles. DM particles are excited via collisions, creating closed Electromagnetic (EM) oscillations about a DM particle, transforming the DM particle into our elementary particles. DM particles are the fundamental particle.

Nordström's theory rivaled Einstein's relativity. Nordström describes the gravitational flux as scalar rather than a vector as treated by Einstein. Nordström's theory failed to mathematically demonstrate photon refraction when passing by stars whereas Einstein's relativity predicted the refraction, which resulted in acceptance of Einstein's theory. However, reevaluation of Nordström's theory is warranted because new observations are correctly predicted by Nordström's theory and not by Einstein's relativity. Examples are: (a) the expanding, accelerating universe, (b) galaxy rotation, (c) variation of the big **G** gravitational constant. The variation of **G** is currently a topic for experiments. [3]

Nordström modifies Kepler's law by defining mass as a function rather than a constant. That function is dependent on the scalar gravitational field, defined as $g(\phi)$.

Accordingly, the kinetic energy equation is modified by adding a gravitational scalar value modifier to the mass, $g(\phi)$, which predicts matter to accelerate at the edge of the universe because of smaller mass value and conservation of energy.

$$K = \tfrac{1}{2}\, g(\phi)\; m\; v^2$$

A photon is theorized to be a vector EM excitation of the gravitational flux. Thus no mass or rest mass, only energy and momentum are transported by a photon on the gravitational field.[2]

A New Concept for Mass

Table Of Contents

Table of Figures

A New Concept for Mass

1. Definition for Mass

This paper proposes a model for mass that combines two theories.

The first theory advanced by Gunnar Nordström, proposed that mass is a scalar result of the gravitational field. Thus mass is a function of the gravitational flux at each particle location and will vary depending on the intensity of the scalar gravitational flux contributions from all surrounding particles.

The second theory predicts that all mass is composed of DM particles, and high velocity collisions at the edges of rotating DM galaxies create our matter. The fundamental elementary particle is a DM particle.

Mass is currently defined by measuring it in two fundamental ways, both using the gravitational field. These are by the gravitational potential between two defined bodies and inertial energy required to obtain a specific velocity. Gravitational potential mass and inertial mass have been shown to be the same.

In both cases mass remains mysterious because it is defined by its measurements in a gravitational field, and mass is presumed constant except by effects of relativity and mass energy conversion ($E=mc^2$).

Mass is also described as a particle, which suggests it has an outer surface. Currently even fields, such as the gravitational field, are considered to interact using particles limited to the speed of light. We have a tautological definition. Mass is defined by the effects in a field and fields are defined as particles which have mass in some form.

Consider a much simpler explanation wherein mass is a result of gravity and dependent on the gravitational field. All matter may be a field.

1.1. Proposed New Theory for a Mass Particle

A mass particle first described herein discusses only the gravitational aspects, not the EM forces also associated with elementary particles. The preferable particle is a DM particle because only the gravitational field is present.

First a new picture of a particle is advanced. Current views of a particle are that it has a surface, which defines the particle. Consider a mass particle being more like a fuzzy bubble wherein some inner radius exists. The mass particle is not a point. A minimum radius inner boundary is required to avoid the gravitational field approaching infinity at the center of the particle. However, there is no outer radius since the gravitational field extends to infinity. Thus our mass particle is best described as a fuzzy bubble as shown in Fig. 1. The particle does not have a size since it does not have a defined outer radius.

The mass particle's gravitational field describes the mass particle. The gravitational field is shown as a dotted pattern with dot density highest near the center and decreasing with radius. Since the fuzzy bubble does not have an outer radius, it is mathematically dimensionless except for the inner boundary. However some effective outer radius does exist.

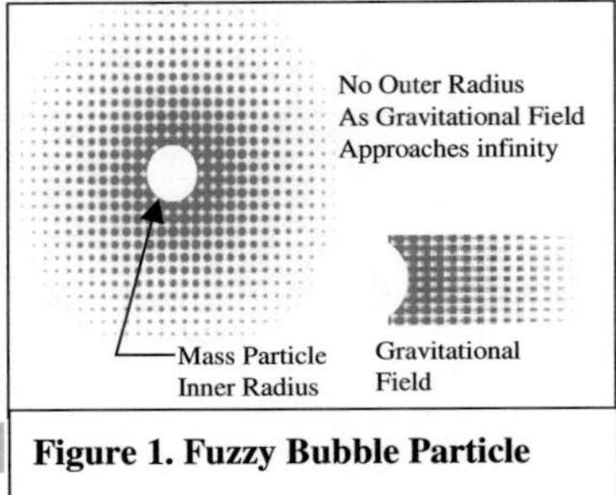

Figure 1. Fuzzy Bubble Particle

The mass particle is quantized, thus mass must be described in quantum mechanics. Some minimum size mass particle may be predicted.

Since the mass particle is described by the gravitational field that indicates that mass is a field, and all mass is a monopole gravitational attractive field. Quantum gravity and the fundamental mass particle may be one in the same.

Wikipedia describes Mass as both a property of a physical body and a measure of its resistance to acceleration when a net force is applied. Wikipedia's website describing mass is: <https://en.wikipedia.org/wiki/Mass>.

1.2. Trading of Potential Energy Between Particles

Envision a gedanken experiment wherein a $mass_1$ particle is inside a spherically shaped distribution of mass elements. Such a sphere is described as the shell theorem.[4] The gravitational attraction on the m_1 particle in the center of the spherical $M_{2,n}$ mass elements is balanced with no net force on m_1 from the $M_{2,n}$ elements. Thus m_1 at the center of the $M_{2,n}$ sphere does not have any net gravitational forces acting to put m_1 in motion because

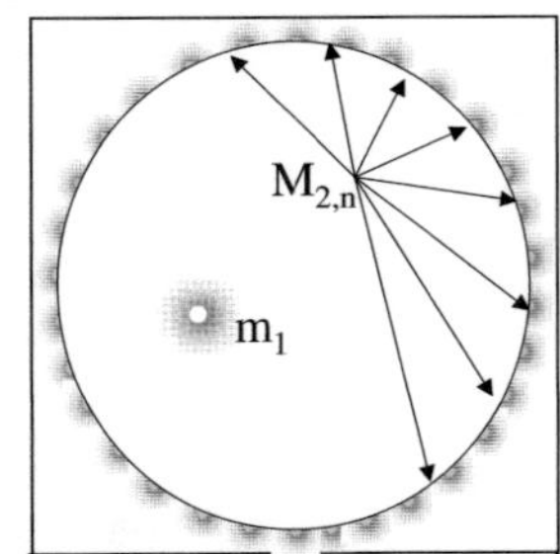

Figure 2 **Gedanken Gravitational Potential Balanced**

in such a field the vector forces equal the vector sum of the forces. The potential energy for m_1 is independent of position inside the sphere, and remains constant.[4] The field around the m_1 particle is simply the vector sum of the fields around each particle element $(M_{2,n})$ making up the sphere. The mass, m_1, has the same potential energy no matter *where* m_1 is inside the sphere. Therefore no force; no work is done when m_1 moves about inside the sphere.

The point of this sphere experiment is to illustrate that the net gravitational force on m_1 remains constant, but all the mass elements must actively trade potential energy between each other in order to maintain equilibrium. Based on this observation, it is theorized that constant communication occurs between all mass elements, and because of the undefined fuzzy outer boundary, all mass elements are basically inside each other. Communication is theorized through the field, not particles as currently theorized. The velocity of this communication is a function of the gravitational field stiffness and may not be associated with photon velocity, thus different physics rules may apply.

1.3. Data Supporting Nordström's Theory

Nordström's prediction is that mass is a direct function of the scalar gravitational field at the mass particle's location. Thus the mass of a specific particle is based on its own scalar gravitational field as well as the sum of all adjacent particles contributing gravitational fields at that location. Mass density, $\rho = g(\phi)$ where $g(\phi)$ is some function to be determined.[1]

Nordström's second theory for a scalar gravitational field addresses all the observations listed below, except refraction, if mass density, ρ, is a function of $g(\phi)$.

1.3.1. Expansion of the Universe

The expansion of the universe has been observed to be accelerating at the outer regions. According to Nordström's theory, mass density, ρ, decreases in the outer universe regions by a function $g(\phi)$.

Conservation of energy requires a velocity increase in order to conserve kinetic energy as calculated in the standard kinetic energy equation with the addition of Nordström's mass modifier g(φ) added. K is the kinetic energy, g(φ) is a scalar sum of the gravitational flux, m is the mass and v is the velocity.

$$K = \tfrac{1}{2}\, g(\phi)\, m\, v^2$$

1.3.2. Gravitational Constant G

Similarly the Gravitational Constant, big **G**, appears to vary. The mass density changes based on proximity of large bodies such as the sun and moon which cause variation of g(φ) and mass density, ρ, further causing the big **G** to vary. [2]

1.3.3. Solar System Rotation

The solar system's rotation does not comply with Kepler's law. DM was theorized to explain the rotation discrepancy. Nordström's theory may explain in part that discrepancy.

The evidence for DM includes more evidence than galaxy rotation, thus Nordström's theory does not replace DM theories.

1.3.4. Photon Refraction Passing Mass Concentrations

Photon attraction to massive bodies caused an apparent gravitational refraction. Relativity is the current accepted theory that describes gravitational refraction. Nordström's mathematical analysis did not predict photon gravitational refraction resulting in abandonment of his theory.

However, experiments show refraction of light in a vacuum using a magnetic field.[5] This data indicates that gravitational flux φ may also have an index of refraction which would predict refraction of photons when passing a star. Mass is not involved except to generate a high gravitational field.

Gunnar Nordström's theory may have been too quickly abandoned because the particular mathematical approach had failed to predict this phenomenon. If the gravitational scalar field also has in index of refraction based on field strength, photon refraction is mathematically described consistent with Nordström's theory.

1.4. Possible Changes to the Conservation Laws

Conservation laws state that a particular measurable property of an isolated physical system does not change as the system evolves over time. Examples of accepted conservation laws are conservation of mass-energy, conservation of linear momentum, conservation of angular momentum, and conservation of electric charge. Wikipedia: https://en.wikipedia.org/wiki/Conservation_law .

Suggested additional/modified conservation laws are:

(1)   Individual mass particle are conserved in number; however, the particle's mass density is a function of scalar gravitational field ϕ. Thus a single mass particle cannot be destroyed or changed to energy.

(2)   Gravitational fields are conserved.

(3)   Energy is conserved but changes energy forms.

(4)   Mass-Energy are not interchangeable, mass is a function of the gravitational scalar field strength.

These conservation laws suggest everything is made up of a mysterious gravitational fuzzy bubble and energy.

2. Dark Matter

Wikipedia describes DM as matter that constitutes approximately 85% of the matter in the universe and about a quarter of its total energy density referred to as Dark Energy (DE). The majority of DM is theorized to be non-baryonic in nature, composed of some unknown subatomic particles. https://en.wikipedia.org/wiki/Dark_matter

2.1.   DM Theorized Particle Family

DM particles were earlier described in Figure 1 as fuzzy bubbles having an inner boundary and an infinite outer boundary, or more precisely no outer edge. A DM particle is described in terms of its gravitational scalar flux, which provides a mass like property. Some minimum size particle is expected to exist, and combination of these minimum particles will create a family of DM particles. These fuzzy bubbles when merging create a family of different size DM particles. The process is shown in Figure 3 similar to a Feynman diagram.

The DM particles have only positive attractive force without a negative force. Thus the complex standard model symmetry is not expected to apply to the DM particle family because of the lack of a negative scalar gravitational field.

The DM particle family is simply a series of DM particles with larger gravitational fields, inner radiuses, and apparent mass quantized based on the smallest DM particle.

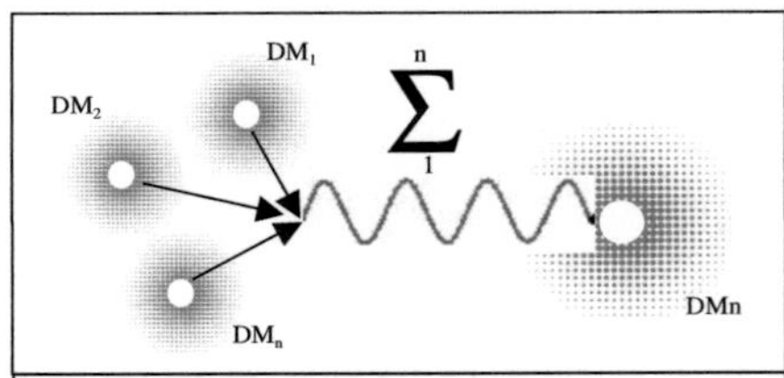

Figure 3 **DM particles combining creating a family of DM particles**

2.2. DM Particles Collect Under Influence Of Gravity

DM particles when collecting under the influence of gravity also concentrate large amounts of Dark Energy wherein the original gravitational potential is transformed into DM particle kinetic energy.

DM particles have the ability to pass through other DM particles and our mass only being affected by the weak gravitational force. Thus the particles in a DM concentration are all in various orbits around their collective center of mass, and pass through each if their respective kinetic energies exceed the weak gravitational attractions.

2.3. Spinning DM Disk Form

Because we observe all galaxies in a disk form, it is theorized that the DM masses first must form into a disk as described above.

Two DM masses can be formed into a disk if two DM masses collide off center. The result is a disk like DM mass wherein much of the DM particles are rotating in the same direction with high velocities as shown in Fig. 4

The Super Massive Black

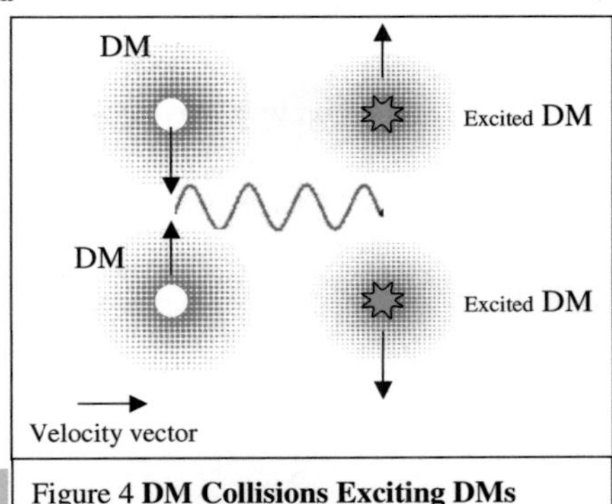

Figure 4 **DM Collisions Exciting DMs Creating Our Matter**

Hole (SMBH) at the center of galaxy NGC 1365 has a disk edge velocity at 84% the speed of light. [6] Such a SMBH, if typical, creates a very high-energy particle accelerator. The rotation rate of inner galaxies has been measured at 84% the speed of light. Thus nature's high-energy accelerator is theorized to create our matter upon collisions between DM particles. See Figure 5.

The proposed theory is that the DM particles when colliding are energized wherein their inner boundary gravitational field becomes excited, causing an Electro Magnetic (EM) field to be generated. Thus the DM particle becomes our mass, when the EM field is generated.

The excitation waveforms can take many forms providing many degrees of freedom. The DM particles of different sizes also provide many degrees of freedom, and the waveforms on these DM particles multiply the degrees of freedom. Many of these excited particles may not be stable. The challenge is to identify an excited DM of a particular size with a specific excitation waveform matched to our current elementary particles.

2.4.   Particle Physics Standard Model

The particle physics Standard Model is based on forces and energy characteristics. These particles are identified as elementary particles. The Standard Model describes three of the four theorized fundamental forces, (the electromagnetic, weak, and strong interactions) and excludes the gravitational force. The Standard Model identifies all known elementary particles. Figure 5 shows the standard model. [7]

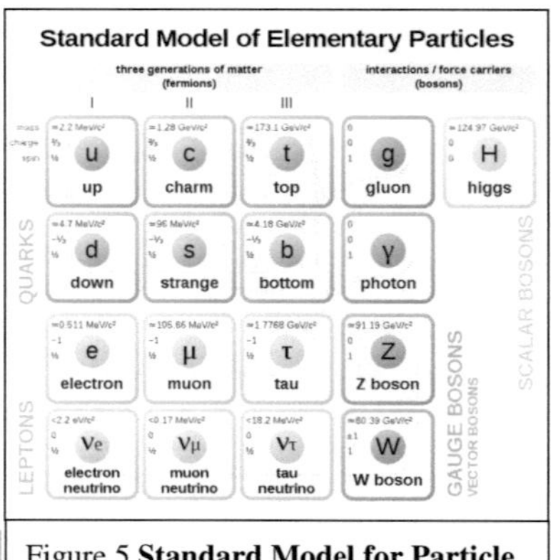

Figure 5 **Standard Model for Particle Physics**

Some elementary particles are defined by their energy characteristics. For example, particle physics describes particles with spin s as a 2s-dimensional spinor field, where s is an integer or a half-integer. Spinor field describes fermions, while a tensor field describes bosons.

2.5. Proposed Particle Physics Model

A proposed model only has three forces; these are the gravitational, electric and magnetic fields. The electric and magnetic fields are coupled when in motion. The gravitational field is a monopole positive field whereas the electric and magnetic fields are positive and negative based on direction.

The fundamental particle is the smallest DM, which may or may not support excitation. The DM family is a series of combined DM particles.

The DM family may be energized with many waveforms all having different stabilities. Each DM particle of a particular size can be energized with a family of excitation waveforms.

The result is a very large number of fundamental particles.

Once a DM particle is excited, the excited particles interact with many new degrees of freedom.

The result is a very large number of elementary particles, too many to classify with names and provide in a table. The particle would need to be identified by its DM number and its waveform. Waveforms introduce spin.

2.6. The Role of Quantum Mechanics

Quantum is integral to physics as described herein because the Fuzzy Bubbles are quantized in number. However, their respective masses are not quantized, but can vary in an analog manner dependent of the scalar gravitational strength. Similarly energy is not quantized except when associated with a wave or a fuzzy bubble. The energy "container" forces a quantization of an energy packet.

Energy forms, such as, gravitational potential and kinetic energy associated with a particles are not quantized. Similarly dimension and time are not quantized and suspected as not variable as relativity predicts.

2.7. Describing the Gravitational Field

2.7.1. Feynman's Description

Feynman describes a "field" as any physical quantity which takes on different values at different points in space.[8]

2.7.2.   Newton's Description

Newton describes the gravitational field g around a single particle of mass M as a vector field whose magnitude at every point is calculated applying the universal law, and represents the force per unit mass on any object at that point in space.

2.7.3.   Nordström's Scalar Measurement

There is also a scalar potential energy per unit mass, Φ,

$$G = F/m = -(d^2R)/dt^2 = -(GM)\,\vec{R}/|R|^2 = -\nabla\phi$$

which is the gravitational potential. The gravitational field equation is:

F is the gravitational force, m is test particle mass, **R** is test particle position, $\vec{R}$ is a unit vector in the direction of **R**, t is time, **G** is the big **G** gravitational constant, and ∇ is the del operator. [1]

Mass may just be a result of the gravitational field, ϕ. That indicates mass is a field, and all matter is a field as theorized by Hobson. [9]

2.7.4.   Faraday's Description of a Field

In 1845 Michael Faraday coined the English term "field" which he described as a space-filling "lines of force" capable of acting on matter. [10]

2.8.   The Photon and the Gravitational Field

Waves transport energy. A wave travels through a medium transporting energy from one location to another location without transporting matter. A photon is an EM field, which transports energy and momentum.

Wikipedia states an "electromagnetic field can possess momentum and energy. A particle creates a field, and a field acts on another particle, and the field has such familiar properties as energy content and momentum, just as particles can have." <https://en.wikipedia.org/wiki/Field_(physics)>. [11]

2.8.1.   The Photon Travels on a Gravitational Field

Consider the opposite wherein a field creates the particle as postulated by Art Hobson[9]. Thus the gravitational field's Fuzzy Bubble is the source of mass.

Consider a photon to be a disturbance of the gravitational field as an open wave. The disturbance creates the EM field in the same manner as described earlier for the Fuzzy Bubble with its closed wave. Thus the Gravitational Field acts in the same manner as the postulated aether, except it is not a fluid but a field. (This must be postulated by an earlier researcher, but could not find this postulate.)

The Scalar gravitational Field has properties of an index of refraction for photons. The photons will slow down in a high intensity field and speed up in a lower intensity field such as near the outer boundary of a universe. The literature suggests such a phenomena a possible explanation for the Red Shift.

3. Conclusion

Our elementary particles are excited fundamental DM particles wherein excitation occurs at the edge of DM disk galaxies.

Matter in the form of Fuzzy Bubbles is conserved, but mass density, ρ, is a function of scalar gravitational field ϕ.

Energy is conserved but is interchangeable in its forms of gravitational potential, kinetic energy and waveforms on the gravitational field, which includes spin. When associated with DM, this energy is the current theorized Dark Energy (DE).

Nordström's theory suggests matter inside massive bodies such as the sun or Black Holes may have properties not observed in our realm of the universe. Similarly at the edges of the universe the same is also true.

As particles exit the universe, mass evaporates into the smallest DM. Thus we live in two universes, an infinite DM universe and a finite elementary particle universe, which is a subset of the infinite DM universe. All matter is composed of fields and energy.

4. References:

1. Wikipedia Nordström's Second Theory Of Gravitation <Https://En.Wikipedia.Org/Wiki/Nordström%27s_Theory_Of_Gravitation> Downloaded 9 Apr. 2019

2. Daniel Lee Stark "Fundamental Particle For All Matter Is Dark Matter" Jan. 30, 2019 Published Academia.Edu Https://Independent.Academia.Edu/Danstark

3. Rupert Sheldrake "How The Universal Gravitational Constant Varies" Downloaded Https://Www.Sheldrake.Org/Essays/How-The-Universal-Gravitational-Constant-Varies> Downloaded 9 Apr., 2019

4.   Wikipedia Shell Theorem For Inside A Sphere Https://En.Wikipedia.Org/Wiki/Shell_Theorem Downloaded 9 Apr., 2019

5. R. Battesti, C. Rizzo, "Magnetic And Electric Properties Of Quantum Vacuum" Physics Optics, 5 Nov., 2012

6. Nasa/Jpl Pasadena, Calif. – Two X-Ray Space Observatories, Nasa's Nuclear Spectroscopic Telescope Array (Nustar) And The European Space Agency's Xmm-Newton. "Nasa's Nustar Helps Solve Riddle Of Black Hole Spin" Nustar, Feb. 27, 2013

7. Wikipedia Standard Model For Elementary Particles <Https://En.Wikipedia.Org/Wiki/Standard_Model> Downloaded Apr. 9, 2019

8.<Http://Www.Feynmanlectures.Caltech.Edu/I_13.Html>Downloaded 9 Apr., 2019

9. Hobson, Art (2013). "There Are No Particles, There Are Only Fields". American Journal Of Physics. 81 (211): 211-223. Arxiv:1204.4616.Bibcode:2013amjph..81..211h. Doi:10.1119/1.4789885.

10. Wikipedia, Https://En.Wikipedia.Org/Wiki/Line_Of_Force> Visited April 9, 2019

11. Wikipedia <Https://En.Wikipedia.Org/Wiki/Field_(Physics)> Downloaded 1 May, 2019

5. Figure Citations

Figure in Abstract Author's own work

Figure 1 Author's own work

Figure 2 Author's own work

Figure 3 Author's own work

Figure 4 Author's own work

Figure 5 Author: Miss MJ., Title. Standard Model of Elementary Particles, Created: 27 June 2006, updated 2008, Fermilab, Office of Science, United States Department of Energy, Particle Data Group, Downloaded Apr. 9, 2019. See Ref. 7

YOUR KNOWLEDGE HAS VALUE

- We will publish your bachelor's and master's thesis, essays and papers

- Your own eBook and book - sold worldwide in all relevant shops

- Earn money with each sale

Upload your text at www.GRIN.com and publish for free